Construction

By Beth Sidel and Karen Redeker-Hicks
Illustrated by Barb Tourtillotte

Published by Totline Publications
an imprint of
McGraw-Hill Children's Publishing

Editorial Director: Hanna Otero
Editor: Joanne Corker
Art Director: Kristin Lock
Graphic Artists: Randy Shinsato, Kathy Kotomaimoce
Cover Illustrator: Barb Tourtillotte

McGraw-Hill
Children's Publishing
A Division of The **McGraw·Hill** Companies

Published by Totline Publications
An imprint of McGraw-Hill Children's Publishing
Copyright © 2002 McGraw-Hill Children's Publishing

Totline Publications grants permission to the individual purchaser to reproduce page 11 in this book for noncommercial, individual, or classroom use only. Reproduction of this page for an entire school system is strictly prohibited. No other part of this publication may be reproduced, stored in a retrieval system, or transmitted, in any form or by any means, electronic, mechanical, photocopying, recording, or otherwise, without the prior written permission of the publisher.

Send all inquiries to:
McGraw-Hill Children's Publishing
3195 Wilson Drive NW
Grand Rapids, Michigan 49544

All Rights Reserved • Printed in the United States of America

Preschool Projects: Construction–grades PreK-K
ISBN: 1-57029-325-2

1 2 3 4 5 6 7 8 9 07 06 05 04 03 02

Contents

Introduction 4
About This Book 4
Concepts 5
Bulletin Boards 6
Learning Centers 6
Community Resources 9
Related Websites 9
Web 10
Parent Letter 11
Calendar 11
Activity 1: Introduction to Architecture 12
Activity 2: Discovery Walk 13
Activity 3: Similarities and Differences 14
Activity 4: Paper-Shape Buildings 15
Activity 5: People in the Trade 16
Activity 6: Exploring Building Materials 17
Activity 7: Marshmallows and Toothpicks 18
Activity 8: Sugar Cubes 19
Activity 9: Clay 20
Activity 10: Balsa Wood 21
Activity 11: Blueprints 22
Activity 12: Elevation Plans 23
Activity 13: Floor Plans 24
Activity 14: Building a Model 25
Activity 15: Architectural Project 26
Activity 16: Class Elevation Plan 27
Activity 17: Class Floor Plan 28
Activity 18: Class Model 29
Activity 19: Getting the Site Ready 30
Activity 20: Final Project 31
Bibliography 32

Introduction

About This Book

Most young children love to build structures! They're intrigued by backhoes, bulldozers, hammers, and drills. They're fascinated by skyscrapers, pyramids, and castles. And they love to see how tall or how wide they can make their own towers, bridges, and tunnels.

Construction is an exciting new curriculum resource designed to capture your children's enthusiasm—and bring it to the classroom. It contains twenty original activities about the construction building process. Each thematic idea has been written by and for early childhood teachers and promotes learning through play. Furthermore, the lessons are sequential, and build upon previous activities and concepts—so that by the end of the month, your children will have learned about construction as a whole rather than in segments.

As you browse the pages, you will quickly see that most of the materials are simple and inexpensive and utilize community and Internet resources. Many of the activities encourage your children to use their imaginations and transform everyday items (e.g., marshmallows, toothpicks, sugar cubes) into all sorts of constructive paraphernalia. The ideas also incorporate challenging activities across the curriculum and include science, math, art, and language.

Before you begin Construction with your children, be sure to examine the introductory tips. These valuable pages include a list of building concepts, creative bulletin-board displays, and original learning-center ideas. You will also find useful community resources and websites, a web for extended exploration, and a helpful parent letter and calendar. At the end of the book there is an extensive bibliography to support and enhance this program in your classroom.

It is hoped that Construction will stimulate and enrich the learning experiences of your young architects. Let's get ready to build!

Concepts

As your children work through the creative activities in this unit, they should learn the following concepts. Be sure to keep this reference list handy and share it with parents and other visitors to your classroom.

Architecture: The People and Process

- Architecture is the process of construction. It is the science and art of designing and erecting buildings and other artificial structures.
- Architects are the people who design buildings and other forms of construction. They are hired by people who need something built.
- Before architects begin drawing, they meet with the owners to find out what they need in their structure.
- Architects draw special building plans called blueprints.
- Architects use special tools to help them draw the building plans.
- Each building has a set of blueprints. These include floor plans, foundation plans, elevation plans, plot plans, and sectional drawings.
- Elevation plans show what the building will look like from the outside.
- Floor plans show what the building will look like from the inside.
- Architects may use the blueprints to make models and help show what the final building will look like.

Construction: The People and Process

- Buildings are constructed from different materials to meet certain requirements (e.g., owner needs, budgets, weather, location, building codes, styles).
- Building materials have their own unique properties.
- Buildings are made up of different shapes. They have features and designs that are similar and different.
- There are many tradespeople involved in the building process. These include contractors, framers, carpenters, plasterers, plumbers, electricians, roofers, tilers, cabinet makers, painters, and masons.
- People who work in construction need a well-thought out plan on paper before they begin to build. They use blueprints to figure out how to build construction projects.
- The building site needs to be prepared before building can begin. This is usually done by using big machinery.
- Different crews of tradespeople work together or in turn to pour foundations, put up roof and wall frames, and finish the exterior walls and roof. They install doors and windows and run electrical, plumbing, gas, and other lines. They also complete and paint the interior walls and ceilings and install many fixtures and furnishings.

Bulletin Boards

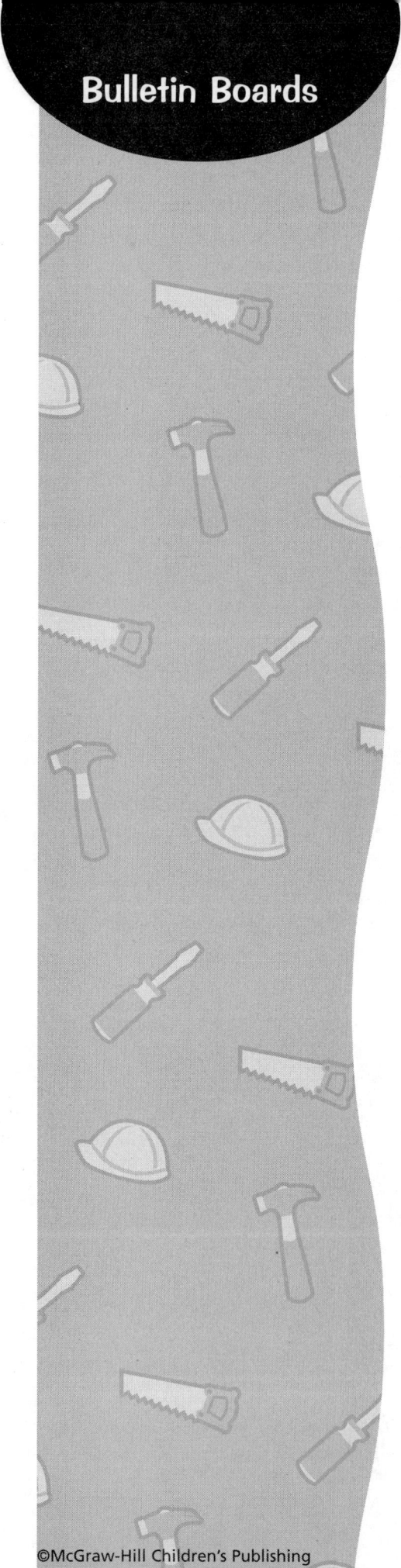

Bulletin Boards

Here are some eye-catching, child-centered suggestions for your classroom bulletin boards. Be sure to refer to your construction displays to visually reinforce the topic with your children.

- Provide each child with paper, crayons or markers, and scissors. Ask the children to close their eyes and imagine that their home is right there in front of them. Have them open their eyes and draw a picture of their residence. When finished, they can cut out their houses and staple them to the bulletin-board. The children may also paint precut paper trees and a road for their "town." Complete the scene with an appropriate title such as "Our Community" or "Where We Live."
- Divide the bulletin board into three sections. For the first part of the scenario, the children can help illustrate an empty construction site. They can glue sand onto paper and staple the "ground" to the board. The children can also paint or color some big precut paper machine shapes (e.g., bulldozers, backhoes) and add these to the site. For the second section, the children can help glue craft sticks onto paper to form a building frame, then attach the structure to the board. For the final section, cut out strips of paper siding and a triangular roof. The children can help paint and staple the pieces onto the board to form a building. They can complete the scene with painted paper trees and a title such as "Building Our School."
- Challenge your young architects to design and build imaginative two- or three-dimensional houses of the future. They can decorate their homes with collage materials and textures (e.g., craft sticks, aluminum foil, sand, straw) and add vehicles, trees, people, pets, and a title such as "Our Homes of the Future."
- Set up an area in your classroom where the children can display their artwork and models from the unit (e.g., paper-shape buildings, marshmallow-and-toothpick structures, sugar-cube buildings, clay homes, balsa-wood structures, blueprints, final class project). The children can add to the display with pictures and information from their own Internet and library-book searches.

Learning Centers

Browse these inexpensive and fun tips for enriching your classroom learning centers. These ideas encourage self-guided play and exploration.

Dramatic Play

- Set up a carpentry table with a variety of simple tools such as hammers, blunt saws, screwdrivers, nails, wood, and safety glasses. Encourage the children to pretend they are carpenters building structures in their "workroom."
- Provide some drafting tables (borrowed or homemade), blueprint or drawing paper (from drafting or art-supply stores), pencils, and small milk cartons. The children can imagine that they are architects drawing blueprints and building models.

- Place some hard hats, fluorescent orange vests, and work boots (from parents or tradespeople), cardboard boxes, large toy construction vehicles, and plastic tools in the center. The children can create their own construction sites. (See page 30.)
- Include a dollhouse with multicultural toy people so that the children can play and develop different construction stories.
- Display the class project (see page 31) and provide time for the children to explore and dramatize stories in their special building.

Art Center
- Place assorted construction-paper shapes, glue, and white paper in the center. Encourage the children to use the shapes to create different buildings. (See page 15.)
- Equip the area with small rectangular sponges, red paint, and white paper. The children can sponge-paint assorted "brick houses."
- Provide the children with clay, small tools and picture books about buildings. They can look for interesting structures and try to make their own clay versions of these places. (See page 20.)
- Place recyclables (e.g., small and large milk cartons, cereal boxes, egg cartons, oatmeal containers), glue, paint, and markers in the center. The children can use them to create and decorate imaginary houses.
- Provide small pieces of balsa wood and glue for the children to build little models of houses, cabins, and other structures. (See page 21.)

Block Area
- Place toy construction vehicles (small or large), multicultural toy people, and blocks in the area. The children can use these resources to create unique buildings and communities.
- Provide a variety of blocks (e.g., wooden, brick, foam, plastic, interlocking, castle), and invite the children to build assorted structures and cities.
- Equip the center with wooden or plastic logs, multicultural toy people, and small vehicles. The children can construct different log villages and towns.
- Place a floor-mat road, small toy vehicles, mini wooden or cardboard buildings, and multicultural toy people in the center. The children can make their own small cities.

Media Table
- Cover the bottom of the media table with sand or dirt and add sugar cubes and small toy construction vehicles. The children can use these resources to organize construction sites and build sugar-cube "houses." (See page 19.)
- Include plastic interlocking blocks and building boards in the center. Encourage the children to create their own original cities and towns.
- Provide the children with dirt, straw, water, and brick molds. They can use these materials to make mud bricks and build imaginative structures. (See page 20.)
- Place some ice cubes in the center and challenge the children to make their own icy buildings.

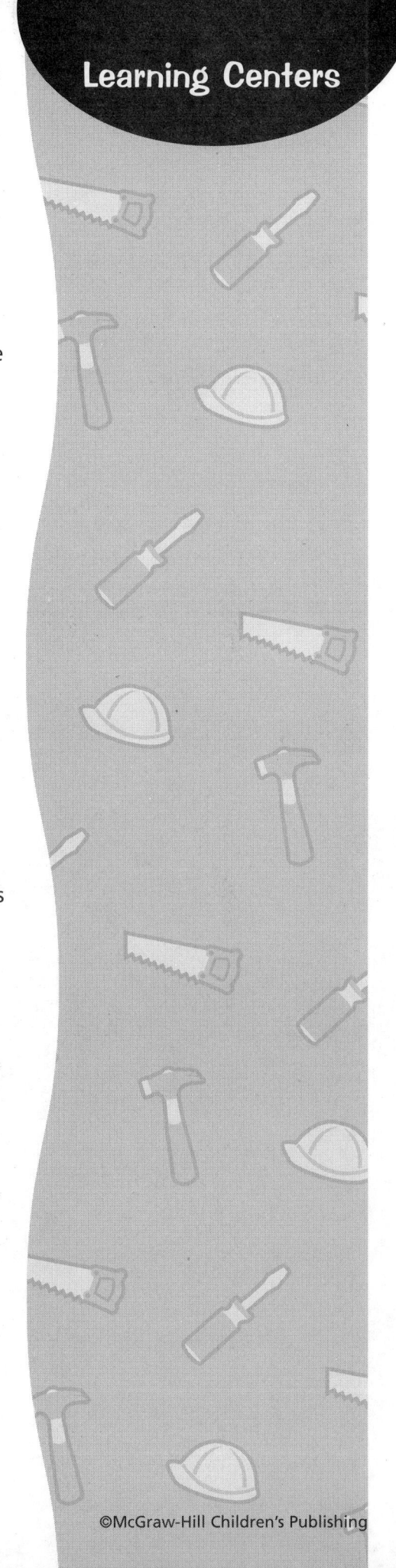

Learning Centers

Learning Centers

Writing/Language Center

- Place blueprint paper (from drafting or art-supply stores), rulers, and pencils in the area. Have the children design and draw their own building plans for all sorts of original and everyday structures. (See page 22.)
- Provide paper and pencils for this partner activity. The children can take turns being the architect or prospective owner. The "owner" can describe the building he or she would like built, and the "architect" can draw the structure on paper.
- Display pictures of items that may be found in buildings (e.g., chairs, tables, cabinets, basins, faucets). Also provide homemade cards printed with the names and outlines of the objects. The children can match the pictures to the words on each card.
- Equip the center with a cassette-tape player, pencils, paper, and assorted geometric shapes (e.g., circles, squares, rectangles, triangles). Record a set of instructions telling the children to trace specific shapes (patterns) and make simple structures. (See page 15.)

Math Center

- Provide pattern blocks and homemade cards with pictures of different block buildings. Challenge the children to use the pattern blocks to recreate these structures.
- Include pictures of different buildings (e.g., houses, hospitals, schools, stores) and invite the children to sort the structures into their different categories.
- Display pieces of dollhouse furniture (e.g., tables, chairs) and homemade floor plans. Have the children match the furniture to the blueprints by correctly placing the pieces on top of the plans.
- Equip the area with highlighters and magazine pictures of buildings. The children can use the highlighters to circle the shapes they find in the structures. (See page 15.)
- Place blueprints (homemade or from architects) and rulers in the center. The children can measure the lengths of the different lines in the plans.

Science Center

- Display a variety of building materials for the children to experiment with and explore. (See page 17.)
- Include several architectural books and pictures showing the cross-sections of different structures. The children can examine the buildings and the way they are designed and built. (See pages 12 and 13.)
- Provide different-sized blocks and homemade cards printed with block-building ideas drawn on them. Invite the children to balance the blocks, using the diagrams as guides. Ask questions such as: "If you put the big block on top of the small block, what will happen?" "What will happen if you move the big block to the right or left?"

Community Resources

Use these valuable community resources to enhance your children's learning, to plan for possible field trips and classroom visitors, and to obtain photographs, videos, and other resources.

- Invite an architect or draftsperson to talk with the children about architecture and to explain his or her role in creating a building.
- Ask architects, draftspersons, and other tradespeople for old blueprints to use in the classroom.
- Have a building contractor or other tradesperson visit the children and talk about his or her role in constructing a building.
- Invite a carpenter or bricklayer to help the children build a small structure in the classroom or playground.
- Visit a construction site and observe the steps involved in building a house, apartment building, store, or other structure. Also look for big machinery at work.
- Visit local historical sites and arrange for a tour of these old buildings.
- Use the parent letter to enlist the support of your children's families. (See page 11.) Complete the thematic calendar with the correct dates before sending it home.

Related Websites

Explore these fascinating websites with your children. Or, send the list home with the parent letter and encourage families to visit these sites together to reinforce classroom learning. They may even like to examine sites for their favorite building blocks and toys.

The Frank Lloyd Wright Foundation was begun in 1940 by Frank Lloyd Wright to promote and support the field of architecture. Visit its website at **www.franklloydwright.org**.

Explore the unusual architecture of the Weisman Art Museum at the University of Minnesota. This amazing building is worth a trip to its website. Go to **www.hudson.acad.umn.edu/surprises**.

The Barn Journal and its website, **www.museum.cl.mse.edu/barn**, is dedicated to the appreciation and preservation of traditional farm architecture. Produced by Michigan State University.

The Great Buildings website is the online gateway to architecture around the world and across history. Look up famous architecture by building, architect, or location at **www.greatbuildings.com**.

Take a virtual tour of the Parthenon or explore the architectural history of the Seven Ancient Wonders at **www.akropolis.net**.

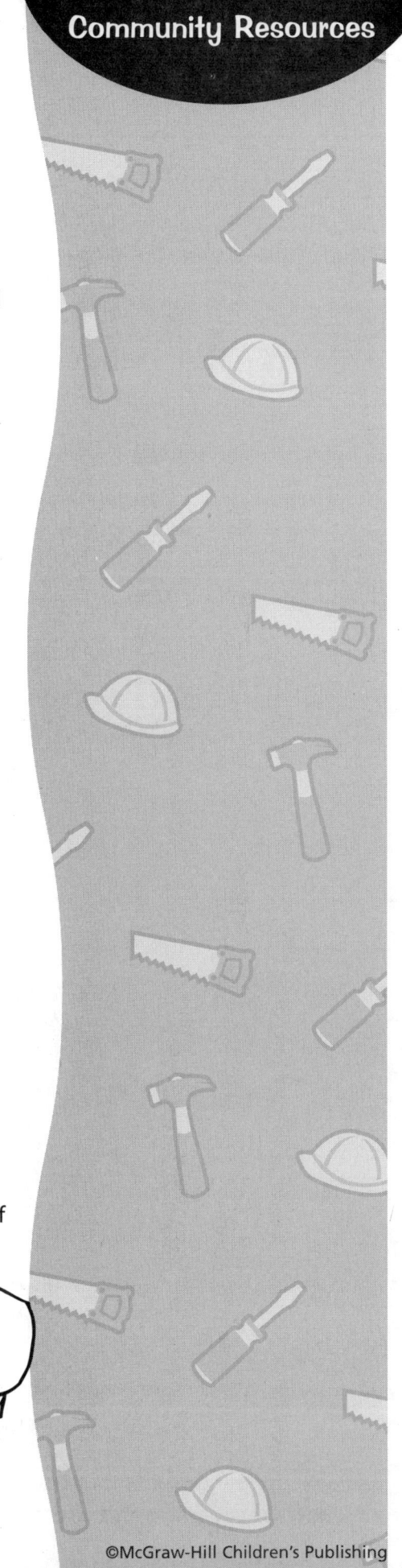

Web

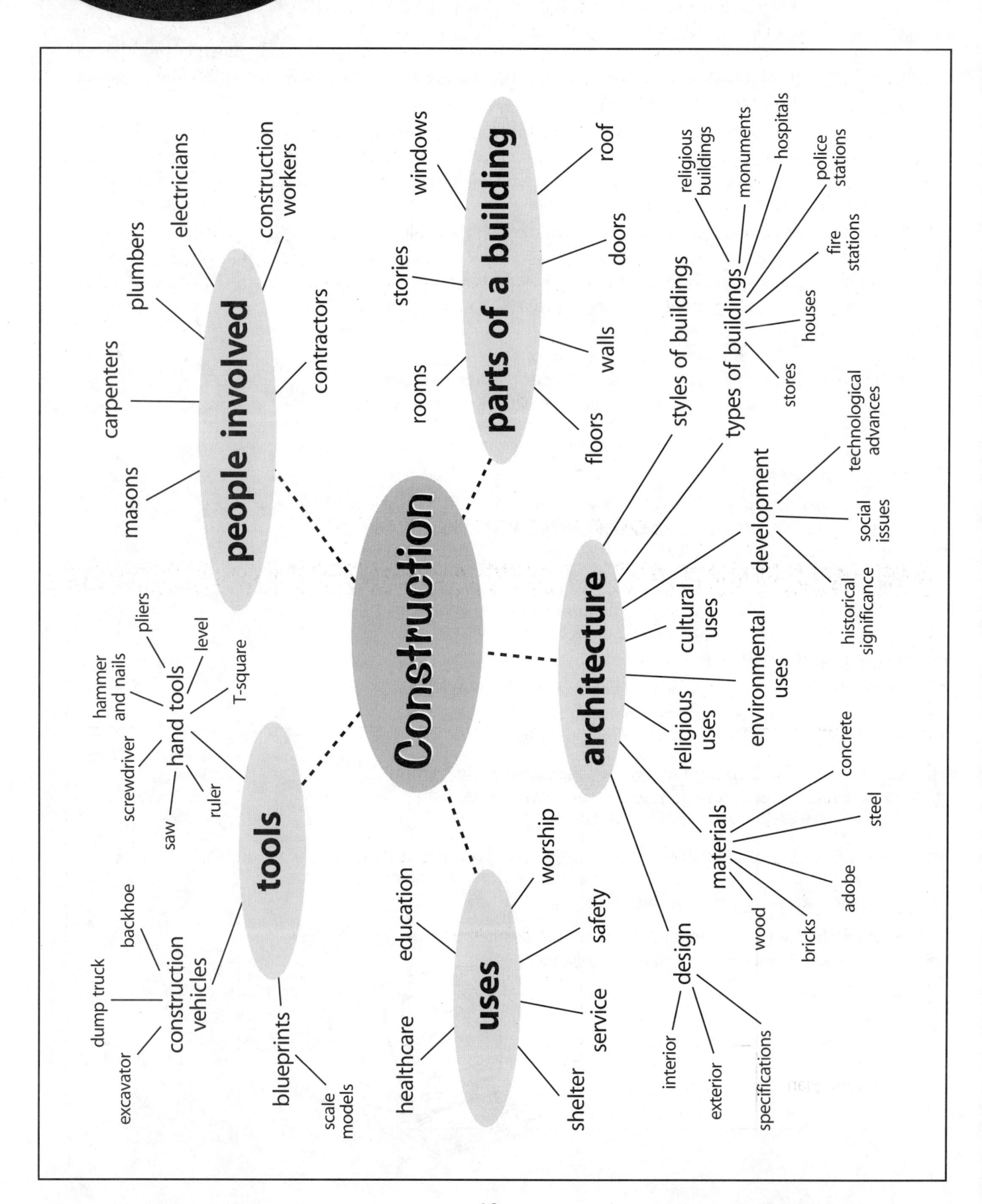

Consult this web of extended ideas if you or your children are looking for additional topics of interest.

Construction

- **people involved**
  - electricians
  - construction workers
  - plumbers
  - contractors
  - carpenters
  - masons

- **parts of a building**
  - windows
  - stories
  - rooms
  - roof
  - doors
  - walls
  - floors

- **architecture**
  - styles of buildings
    - religious buildings
  - types of buildings
    - monuments
    - hospitals
    - police stations
    - fire stations
    - houses
    - stores
  - development
    - technological advances
    - social issues
    - historical significance
  - cultural uses
  - environmental uses
  - religious uses
  - materials
    - concrete
    - steel
    - adobe
    - bricks
    - wood
  - design
    - interior
    - exterior
    - specifications

- **tools**
  - hand tools
    - pliers
    - level
    - T-square
    - hammer and nails
    - screwdriver
    - saw
    - ruler
  - construction vehicles
    - dump truck
    - backhoe
    - excavator
  - blueprints
  - scale models

- **uses**
  - worship
  - safety
  - education
  - healthcare
  - shelter
  - service

Dear Parents,

This month we are very excited to be studying all about construction. Your children will learn about all sorts of different buildings, materials, and tradespeople. They will also learn about the design process by exploring and making different types of blueprints. At the end of the month, the children will use what they have learned to design and create a special class architectural project.

We encourage you and your child to explore buildings and other structures in your neighborhood, including your own home. We also invite you to come into the classroom at the end of the month to view all the wonderful projects your children have worked on throughout this unit.

At the bottom of this letter, you will see a Construction calendar. You can use this outline to help you and your children keep track of the building process. Please remember to ask your "architects" about their daily activities. You may even like to explore some interesting websites together to further enrich their building experiences.

Happy learning!

Sincerely,

Construction Calendar

Monday	Tuesday	Wednesday	Thursday	Friday
Introduction to Architecture	Discovery Walk	Similarities and Differences	Paper-Shape Buildings	People in the Trade
Exploring Building Materials	Marshmallows and Toothpicks	Sugar Cubes	Clay	Balsa Wood
Blueprints	Elevation Plans	Floor Plans	Building a Model	Architectural Project
Class Elevation Plan	Class Floor Plan	Class Model	Getting the Site Ready	Final Project

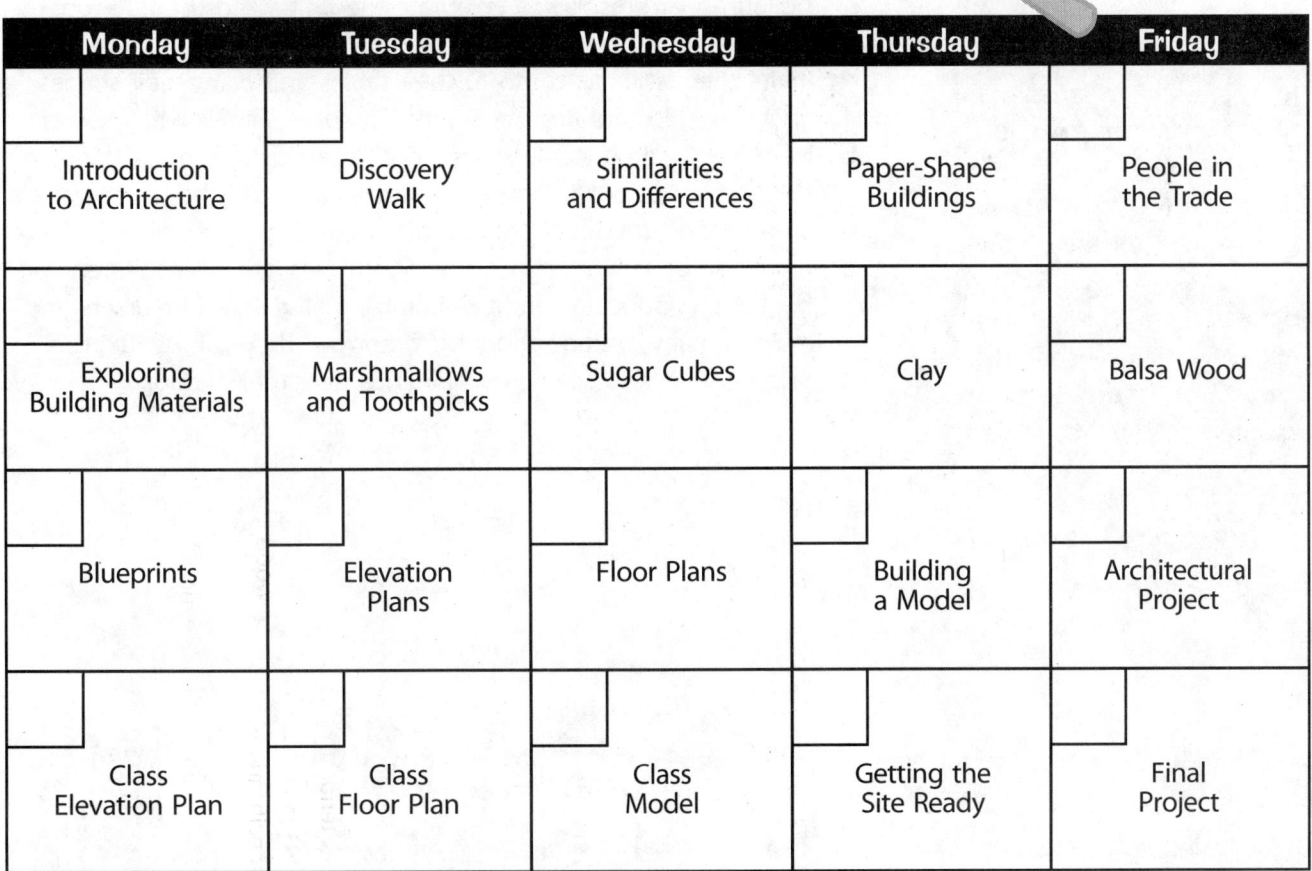

Activity 1

Introduction to Architecture

Objective
The children will learn the meaning of architecture and architect.

Directions
1. Gather the children together and tell them that this is a very special month in their classroom. Show the pictures and models of different architectural structures. Examine them together and ask questions such as: *Do you know what a building is? What buildings do you know? Do you know what architects do? Who helps build structures?*
2. Share the following information with the children.
   - Architecture is the process of construction or building. It is the science and art of designing and erecting structures.
   - Architects design houses, malls, stores, schools, communities, and other buildings. They take many classes to learn how to design and build different structures.
   - Architects are hired by people who want something built. They meet with the owners to find out what they need. Then they design and draw special building plans called blueprints. They may use draftspersons to draw the final plans.
3. Explain that today the children are going to construct a wonderful bulletin-board display to celebrate the unit. Read the title together, and talk about the paper shapes.
4. Invite the architects to go to their tables and paint their shapes. They can decorate the multicultural people shapes with markers.
5. Help the children staple the shapes together to form buildings on the board. Cut out doors and windows and add the paper people around the structures. Discuss the display.
6. Conclude by talking about this month's lessons. The children will be learning about different buildings, materials, and workers and making many exciting models. Tomorrow, they will be taking a discovery walk to see some buildings near their school.

Materials
- Pictures of different architectural structures
- Toy buildings (e.g., dollhouses, forts, castles, towers, bridges)
- Construction paper (white and skin-toned)
- Scissors
- Paint
- Brushes
- Markers

Preparation
- Cut out paper squares, triangles, and rectangles to make into buildings for your bulletin board. Also cut out some people shapes from the skin-toned paper. Make a title for the display such as "Building" or "Construction."
- At each table, place the paper shapes, paint, brushes, and markers for the children.

Teaching Tip
- Although draftspersons also prepare plans, use the term architect to avoid confusion.

Discovery Walk

Activity 2

Objective
The children will notice the details of the buildings found near the school.

Directions
1. Gather the children together and talk about yesterday's activity. Explain that today they are going to take a discovery walk to see the buildings near their school.
2. Ask the children to close their eyes and think about the different structures they see on their way to school. Have each child name one building he or she sees.
3. Invite the children to line up in pairs at the door. Remind them that they need to stay with their partners and listen to their teachers as they take this discovery walk.
4. Once outside, stop in front of each building and ask questions such as: *What is this building? What color is it? What materials do you think it is made out of? Is this building bigger or smaller than the last one? How many floors do you think it has? How many windows? How many doors? Are there any steps? Are there trees, bushes, or flowers around this structure? What other details do you notice?* Videotape or photograph the buildings, if possible.
5. At the end of your walk, ask the children to take turns naming their favorite buildings, and why. Write their responses on the easel and read them together. Ask questions such as: *Which buildings did you like the most? The least? Why?* Review the videotape or photographs, if possible.
6. Conclude by talking about tomorrow's lesson. The children will be drawing pictures of their favorite buildings and looking for similarities and differences.

Materials
- Paper
- Easel
- Tape
- Marker
- Camcorder or camera (optional)
- Videotape or film (optional)
- Television/VCR (optional)
- Photo album (optional)

Preparation
- Tape a length of paper to the easel.

Teaching Tip
- Enlist adult or parent volunteers to accompany the children on the walk and to videotape or photograph the buildings.
- Compile the photographs into an album and add captions and sentences for the children to read together.
- Place a television/VCR in a corner with the Discovery-Walk tape for the children to share and review together or independently.

Activity 3

Similarities and Differences

Objective
The children will learn that buildings have some features that are similar and others that are different.

Directions
1. Gather the children together and talk about yesterday's activity. Explain that today they are going to think about the ways buildings are similar and different.
2. Have the children close their eyes and think about their favorite structures from the previous day. Ask questions such as: *How many doors does your building have? How many windows? How many floors? What color is it? Does it have any special features such as columns, chimneys, or porches?*
3. Invite the children to go to their tables and draw a picture of their favorite building with crayons or markers. If they need help, they may review the videotape or photographs of the walk, if desired.
4. After the children have finished, label their pictures with words such as "door," "window," and "roof," and titles such as "My School," "The Pet Shop," or "Our Library."
5. Have the children return to the circle and share their completed pictures. Ask questions such as: *Who drew a building with one floor? Who drew one with two floors? Three floors? Who drew a building with more than one door? Who drew a school? A house? An apartment building? A store? A skyscraper?* Talk about the other ways these buildings are similar and different.
6. Display the children's artwork on the bulletin board and add a heading such as: "Same or Different?" or "What Is the Same? What Is Different?"
7. Conclude by talking about tomorrow's lesson. The children will be making their own paper-shape buildings.

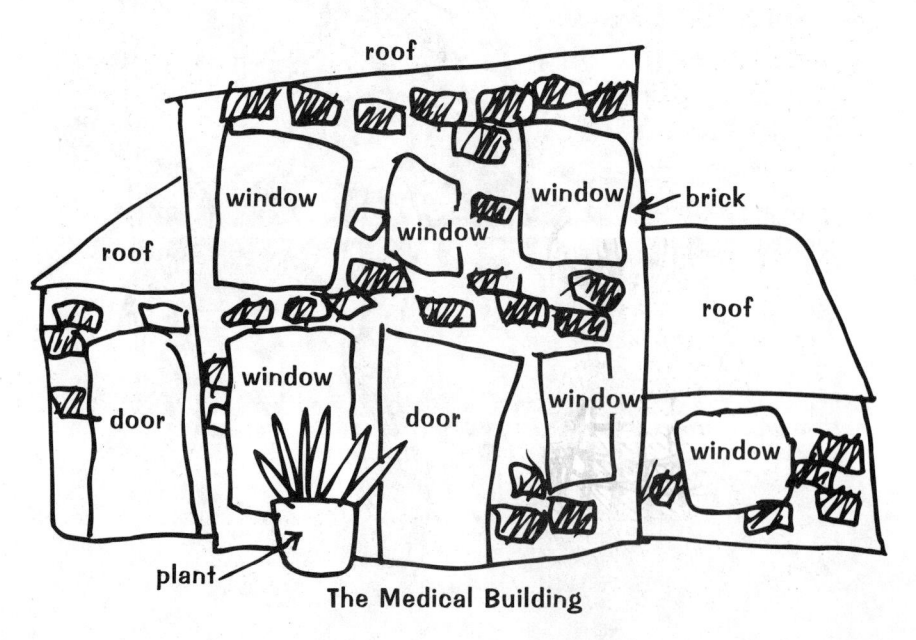

Materials
- Construction paper
- Crayons or markers
- Discovery-Walk tape or photographs (optional) (See page 13.)
- Television/VCR (optional)

Preparation
- At each table, place paper and crayons or markers for the children.

Teaching Tips
- Encourage the children to explore the construction activities in the learning centers around the classroom. (See pages 6–8.)
- Invite the children to examine books in the classroom to find pictures of buildings that are similar to or different from their pictures.

The Medical Building

Paper-Shape Buildings

Activity 4

Objective
The children will learn that buildings are composed of different shapes such as squares, triangles, circles, and rectangles.

Directions
1. Gather the children together and talk about yesterday's activity. Explain that today they are going to talk about the different shapes they see in buildings and make their own paper-shape structures.
2. Show a variety of simple shapes (e.g., squares, circles, triangles, rectangles, hearts), and have the children identify them. Ask questions such as: *Do you see any of these shapes around you? Did you see any of these shapes in the buildings on our walk? Which ones did you see? Where did you see them?*
3. Invite the children to go to their tables and use the paper shapes to make their own buildings. They can glue their structures on construction paper.
4. Have the children return to the circle and share their completed pictures. Ask questions such as: *Did you use all of your shapes to make your building? Do all of our pictures look the same? How are our buildings similar? How are they different?*
5. Display the children's artwork on the wall with a heading such as: "Paper-Shape Buildings" or "Our Buildings Have LOTS of Shapes."
6. Conclude by talking about tomorrow's lesson. The children will be learning about the people who help build a structure.

Materials
- Large and small shape patterns (squares, triangles, rectangles, circles)
- Construction paper
- Scissors
- White glue
- Glue brushes

Preparation
- Use the shape patterns to cut out large and small shapes from colored construction paper. Prepare enough shapes so that each child can have two of each large shape and three of each small shape.
- At each table, place the paper shapes, construction paper, glue, and brushes for the children.

Teaching Tip
- Encourage the children to examine their Discovery-Walk pictures (see page 14) and see how many shapes they can find in their buildings.

Activity 5

People in the Trade

Objective
The children will learn that many different people and tools help build structures.

Directions
1. Gather the children together and talk about today's activities. Explain that they are going to learn that many different people and tools help build structures.
2. Examine the pictures of people working in construction, and the tools they are using. Ask questions such as: *Have you ever seen these people working? Where? What are they doing? What tools do they use? Why?*
3. Share the following information with the children.
   - Many tradespeople are involved in building. These workers include contractors, framers, roofers, tilers, carpenters, plasterers, painters, masons, plumbers, electricians, carpet layers, cabinet makers, and landscapers.
   - Tradespeople use blueprints to figure out how to build construction projects.
4. Read *Building a House* together. Discuss the book and ask questions such as: *What is the story about? What happens first? What kinds of machines are used to get the land ready? What tools do the builders use? What does the cement truck do? What work do bricklayers do? What tools do they use? What work do carpenters do? What about plumbers? Electricians? Painters? What happens when the workers leave? What was your favorite part of the story? Why?*
5. Invite the children to role play or dramatize different tradespeople, and have their classmates identify their roles or occupations. The children may even wear hard hats and orange vests for their dramatizations.
6. Conclude by talking about next week's lessons. The children will be exploring some exciting building materials.

Materials
- Pictures of tradespeople at work (e.g., plumbers, carpenters, masons)
- Tools (real, toy, or pictures)
- Easel (optional)
- Tape
- *Building a House* by Byron Barton
- Hard hats (optional)
- Fluorescent orange vests (optional)

Preparation
- Display pictures of tradespeople and their tools on the easel or around the room. Place real or toy tools nearby.

Teaching Tips
- Invite a tradesperson to visit and talk about his or her work and tools. (See page 9.)
- Visit a building site to observe and photograph construction workers, tools, and machines. (See page 9.)

Exploring Building Materials

Activity 6

Objective
The children will learn that buildings can be made from many different materials.

Directions
1. Gather the children together and talk about last week's activities. Explain that today they will learn that buildings can be created from many different materials.
2. Examine the pictures of structures made from assorted materials and share the following information with the children.
   - Buildings are constructed from different materials to meet certain requirements (e.g., owner needs, budgets, weather, location, building codes, styles).
   - Building materials have their own unique properties. For example, many people build homes from brick because it is durable and can stand up to adverse weather. People living in hot, dry climates may use adobe (dried mud) because it lasts well and helps keep out the heat. People living in colder climates may use wood because it is waterproof and keeps them warm.
3. Read *This Is My House* together, and encourage the children to look for the different construction materials. Ask questions such as: *What materials are used to build houses? What about a barn? A skyscraper? An adobe house? An igloo? What materials are used to build your home?* Write the children's responses on the easel.
4. Show the texture board and examine the samples together. Ask questions such as: *What do you think it would be like to try to build with ice? What about brick? Mud? Straw? What would you use to build your ideal structure?*
5. Conclude by talking about tomorrow's lesson. The children will be building marshmallow-and-toothpick structures.

Materials
- Paper
- Easel
- Tape
- Samples of building materials (e.g., wood, adobe, concrete, brick)
- Cardboard or wood
- Pictures of buildings made from different materials (e.g., concrete, brick, adobe, wood)
- *This Is My House* by Arthur Dorros
- Markers

Preparation
- Tape a length of paper to the easel.
- Prepare a "texture board" by attaching samples of building materials to a piece of cardboard or wood.

Teaching Tip
- Encourage the children to explore architectural and construction websites with their families and share their findings with their classmates. (See page 9.)

WPH39005 *Preschool Projects: Construction* ©McGraw-Hill Children's Publishing

Activity 7

Marshmallows and Toothpicks

Objective
The children will learn about and practice building with different materials.

Directions
1. Gather the children together and talk about yesterday's activity. Explain that today they are going to build with marshmallows and toothpicks.
2. Show the pictures of the building frames to the children. Tell them that making structures with marshmallows and toothpicks is similar to working with real frames.
3. Demonstrate how to build with marshmallows and toothpicks, then ask the children if they think it will be easy or difficult to work with these materials. Talk about some different structures they may decide to make such as miniature dog houses, play houses, candy stores, or houses.
4. Invite the children to go to their tables and freely explore ways to build with these resources. When ready, they can make their structures on the cardboard squares. If they need help, they may ask a friend to help them build.
5. After the children have finished their marshmallow-and-toothpick buildings, write their names on their cardboard bases. Talk about how it felt to build with these materials and provide time for the children to walk around and admire their friends' models.
6. Place the marshmallow-and-toothpick structures in a safe place to harden. Display the completed structures and pictures of building frames for everyone to see.
7. Conclude by talking about tomorrow's lesson. The children will be building sugar-cube structures.

Materials
- Miniature marshmallows (1 to 2 cups per child)
- Toothpicks (½ to 1 box per child)
- Cardboard cut into 10-inch squares (one per child)
- Pictures of building frames

Preparation
- At each table, place marshmallows, toothpicks, and cardboard squares for the children.

Teaching Tips
- The size of each child's structure may depend on the number of children in your class and the time available for this activity.
- Enlist parent or other adult volunteers to help the children build their models.
- Provide each child with a small cardboard or construction-paper blueprint to guide their efforts and help them make their foundations an appropriate size. They may even build their models on their blueprints.

Sugar Cubes

Activity 8

Objective
The children will learn about and practice building with different materials.

Directions
1. Gather the children together and talk about yesterday's activity. Explain that today they are going to build with sugar cubes.
2. Show the pictures of the brick buildings to the children. Tell them that making sugar-cube structures is similar to working with real bricks. (Both are stacked and cemented.)
3. Demonstrate how to build with sugar cubes and glue, then ask the children if they think it will be easy or difficult to work with these materials. Talk about some different structures they may decide to make such as schoolhouses, libraries, or houses.
4. Invite the children to go to their tables and freely explore ways to build with the sugar cubes and glue. When ready, they can make their structures on the cardboard squares. If they need help, they may ask a friend to help them build.
5. After the children have finished their sugar-cube buildings, write their names on their cardboard bases. Talk about how it felt to build with these materials and provide time for the children to walk around and admire their friends' models.
6. Place the sugar-cube buildings in a safe place to dry. Display the completed structures and pictures of brick buildings for everyone to see.
7. Conclude by talking about tomorrow's lesson. The children will be building clay structures.

Materials
- Sugar cubes (80 to 100 cubes per child)
- White glue
- Small plastic containers
- Glue brushes
- Cardboard cut into 10-inch squares (one per child)
- Pictures of brick buildings

Preparation
- At each table, place sugar cubes, glue, brushes, and cardboard squares for the children.

Teaching Tips
- Enlist parent or other adult volunteers to help the children build their models.
- Pour white glue into small plastic containers. The children can use the brushes to apply the glue to the sugar cubes as needed.
- Provide each child with a small cardboard or construction-paper blueprint to guide their efforts and help them make their foundations an appropriate size. They may even build their models on their blueprints.

Activity 9

Clay

Objective
The children will learn about and practice building with different materials.

Directions
1. Gather the children together and talk about yesterday's activity. Explain that today they are going to build with clay.
2. Show the pictures of adobe buildings to the children. Tell them that making clay structures is similar to working with mud and share the following information with them.
   - Adobe is mud that has been mixed with straw or grass, and then formed into bricks. It is dried in the sun to become hard just like clay.
3. Demonstrate how to build with clay, then ask the children if they think it will be easy or difficult to work with this material. Talk about some different structures they may decide to make such as toad houses, cookie shops, churches, or their own homes.
4. Invite the children to go to their tables and freely explore ways to build with the clay. When ready, they can make their structures on the cardboard squares. If they need help, they may ask a friend to help them build.
5. After the children have finished their clay buildings, write their names on their cardboard bases. Talk about how it felt to build with the clay and provide time for the children to walk around and admire their friends' models.
6. Place the clay structures in a safe place to dry. Display the completed structures and pictures of adobe buildings for everyone to see.
7. Conclude by talking about tomorrow's lesson. The children will be building balsa-wood structures.

Materials
- Clay (large lump per child)
- Cardboard cut into 10-inch squares (one per child)
- Pictures of adobe buildings

Preparation
- At each table, place lumps of clay and cardboard squares for the children.

Teaching Tips
- Enlist parent or other adult volunteers to help the children build their structures.
- Point out that the children will be using clay rather than mud to make construction easier for them.
- Provide each child with a small cardboard or construction-paper blueprint to guide their efforts, and help them make their foundations an appropriate size. They may even build their models on their blueprints.

Balsa Wood

Activity 10

Objective
The children will learn about and practice building with different materials.

Directions
1. Gather the children together and talk about yesterday's activity. Explain that today they are going to build with balsa wood.
2. Show the pictures of wooden buildings to the children and share the following information with them.
   - Wooden buildings are usually joined together by nails, pegs, screws, or other fittings.
3. Demonstrate how to build with balsa wood and glue, then ask the children if they think it will be easy or difficult to work with the wood. Talk about some different structures they may decide to make such as frontier cabins, bug houses, museums, or houses.
4. Invite the children to go to their tables and freely explore ways to build with the wood and glue. When ready, they can make their structures on the cardboard squares. If they need help, they may ask a friend to help them build.
5. After the children have finished their balsa-wood buildings, write their names on their cardboard bases. Talk about how it felt to build with the wood and provide time for the children to walk around and admire their friends' models.
6. Place the balsa-wood structures in a safe place to dry. Display the completed structures and pictures of wooden buildings for everyone to see.
7. Conclude by talking about next week's lessons. The children will be learning about blueprints or building plans.

Materials
- Balsa wood (assorted sizes)
- White glue
- Glue brushes
- Cardboard cut into 10-inch squares (one per child)
- Pictures of wooden buildings

Preparation
- At each table, place pieces of balsa wood, glue, brushes, and cardboard squares for the children.

Teaching Tips
- Enlist parent or other adult volunteers to help the children build their structures.
- Point out that the children will be using glue rather than nails or screws to make construction easier for them.
- Provide each child with a small cardboard or construction-paper blueprint to guide their efforts and help them make their foundations an appropriate size. They may even build their models on their blueprints.

Activity 11

Blueprints

Objective
The children will learn about different blueprints and the tools used to draw them.

Directions
1. Gather the children together and talk about last week's activities. Explain that today they are going to learn how a building is designed.
2. Display the building plans (blueprints) and tools. Ask questions such as: *Do you know what these are? What do they look like? Do you know what they are for?*
3. Share the following information with the children as you examine the prints together.
   - Originally, building plans or blueprints were drawn on paper that was treated to make the paper blue and the lines white. Blueprints now refer to copies of building plans.
   - Each building has a set of plans that may include:
     - Elevation plans (show the finished building from the front, rear, and sides)
     - Floor plans (show inside details such as rooms, doors, and electricity outlets)
     - Foundation plans (show the supports and concrete)
     - Plot plans (show where the building should be located on the lot).
     - Sectional drawings (show vertical perspectives such as the roof and floor)
4. Demonstrate how to use a pencil, ruler, and T-square to draw lines and angles on paper to make a building plan. Share the following information with the children.
   - Architects or draftspersons use drawing tools such as pencils with hard leads, standard rulers (12 or 18 inches), and T-squares to make parallel and vertical lines and right angles. They also use computer programs to draw their plans.
   - Architects make sketches on plain or quadruled paper, and often use scales of ¼ inch or ⅛ inch to represent 1 foot. They use blueprint paper, vellum, and other high-quality semi-transparent paper for the final drawings.
5. Invite the children to go to their tables and examine the drawing tools. They can practice making lines and angles on paper to create their own plans.
6. Conclude by talking about tomorrow's lesson. The children will be using the drawing tools to make their own blueprints.

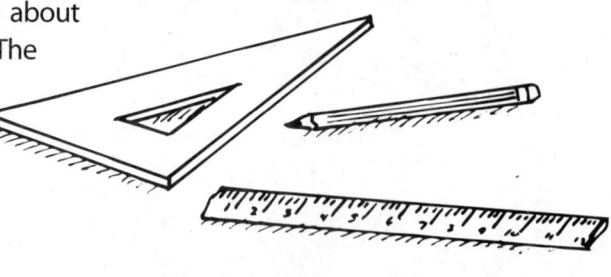

Materials
- Blueprints of floor, foundation, elevation, sectional, and plot plans
- Easel
- Tape
- Paper
- Drawing tools (e.g., pencils, rulers, T-squares)

Preparation
- Tape the blueprints to the easel.
- At each table, place paper and drawing tools for the children.

Teaching Tip
- Sample blueprints may be homemade or donated by architects, builders, or parents.

Elevation Plans

Activity 12

Objective
The children will design and draw an elevation plan for a building.

Directions
1. Gather the children together and talk about yesterday's activity. Explain that today they are going to become architects and design and draw elevation plans.
2. Show the elevation blueprints and examine them together. Ask questions such as: *What do you see on these plans? Do you see anything you recognize? What makes the plans special? Do you think they would be hard or easy to draw? Why?*
3. Challenge the children to think of a building that they would like to build. Have them close their eyes and imagine what the outside of their building would look like. Ask questions such as: *What shape will your building be? Will it be tall or wide? Will it have lots of windows or doors? What materials do you want your building to be made out of?*
4. Invite the children to go to their tables and explain that they are going to use paper, pencils, and rulers to draw what their buildings will look like from the outside. When ready, they can begin working on their elevation plans, refering to the display pictures as necessary.
5. After the children have finished their blueprints, write their names on their plans. Ask questions such as: *Are you happy with your elevation plan? What do you like best about it? How is your plan similar to regular pictures of buildings? How is it different?*
6. Provide time for the children to share their elevation plans with their fellow architects. Then help the children roll up their blueprints and store them in a safe place for future activities.
7. Conclude by talking about tomorrow's lesson. The children will be drawing the floor plans for their buildings.

Materials
- Elevation plans (samples)
- Easel
- Tape
- Pictures of buildings (e.g., houses, skyscrapers)
- Plain paper
- Pencils
- Rulers
- T-squares

Preparation
- Tape elevation plans to the easel and display pictures of buildings nearby.
- At each table, place paper, pencils, rulers, and T-squares for the children.

Teaching Tips
- Sample blueprints may be homemade or donated by architects, builders, or parents.
- Encourage the children to take their time drawing their elevation plans.

WPH39005 *Preschool Projects: Construction* ©McGraw-Hill Children's Publishing

Activity 13

Floor Plans

Objective
The children will design and draw a floor plan for a building.

Directions
1. Gather the children together and talk about yesterday's activity. Explain that today they are going to become architects again and draw and design floor plans for the buildings they drew yesterday.
2. Show the children floor plans and examine them together. Ask questions such as: *What do you see on these plans? Do you see anything you recognize? What makes them special? Do you think these plans would be hard or easy to draw? Why?*
3. Invite the children to go to their tables. Pass out their elevation blueprints and challenge them to imagine what the inside of their building would look like. Ask questions such as: *How many rooms will your building have? Will your building have big rooms or little ones? Will your building have closets? How many? Will your building have windows? How many?*
4. When ready, the children can use the paper and drawing tools to draw their floor plans. They can refer to their elevation blueprints and the sample floor plans as necessary.
5. After the children have finished their blueprints, write their names on their plans. Ask questions such as: *Are you happy with your floor plan? What do you like best about it? How is your floor plan similar to your elevation plan? How is it different?*
6. Provide time for the children to share their floor plans with their fellow architects. Then help the children roll up their blueprints and store them in a safe place for their next activity.
7. Conclude by talking about tomorrow's lesson. The children will be using both of their blueprints to build a model of their building.

Materials
- Floor plans (samples)
- Easel
- Tape
- Children's elevations plans (See page 23.)
- Plain paper
- Pencils
- Rulers
- T-squares

Preparation
- Tape floor plans to the easel.
- At each table, place paper, pencils, rulers, and T-squares for the children.

Teaching Tips
- Sample blueprints may be homemade or donated by architects, builders, or parents.
- Encourage the children to take their time drawing their floor plans.

Building a Model

Activity 14

Objective
The children will learn how an architect uses a blueprint to create a model.

Directions
1. Gather the children together and talk about yesterday's activity. Explain that today they are going to use their elevation and floor plans to build models from balsa wood, cardboard, and glue.
2. Share the following information with the children.
   - After architects complete their blueprints, they sometimes make cardboard or balsa-wood models to show what the final building will look like. They use the blueprints to guide their work, but usually do not need to make actual scale models with all the correct measurements. They may use a scale where 1/4 inch or 1/8 inch of the model equals 1 foot of the real building.
3. Invite the children to go to their tables. Pass out their blueprints and provide time for them to explore the materials and discover ways to build their models.
4. When the children are ready, invite them to make their structures on the cardboard squares. If they need help, they may ask a friend to help them build.
5. After the children have finished their models, write their names on their cardboard bases. Ask questions such as: *How did it feel to build your model? Are you happy with your work? What do you like best about it? How is your model similar to your floor plan? How is it different?*
6. Provide time for the children to walk around and admire the models of their fellow architects. Encourage them to talk about their work with their friends.
7. Place the buildings in a safe place to dry. Display the completed structures and blueprints around the room for everyone to see.
8. Conclude by talking about tomorrow's lessons. The children will be designing a whole-class building.

Materials
- Balsa wood (assorted sizes)
- Assorted pieces of cardboard
- Cardboard cut into 10-inch squares (one per child)
- White glue
- Glue brushes
- Scissors
- Children's blueprints (See pages 23 and 24.)

Preparation
- At each table, place pieces of balsa wood and cardboard, cardboard squares, glue, brushes, and scissors for the children.

Teaching Tip
- Encourage the children to explore building in the block area.

Activity 15

Architectural Project

Objective
The children will experience being part of a classroom building project.

Directions
1. Gather the children together and talk about the previous activities. Ask review questions such as: *What is architecture? What do architects do? What is a blueprint? What types of materials can be used to build with? Who is involved in the building process? What types of tools and machinery are used to help workers build structures?* Discuss the children's responses to refresh their memories.
2. Explain that today they are going to begin working on their class construction project. They will all be the owners, architects, and construction workers. Share the following ground rules to ensure the success of this activity.
   - The children will pretend to get the classroom "site" ready for building.
   - Most materials and tools will be imitation or pretend for reasons of safety, time, and space.
   - The building will only have one floor and an imaginary roof for reasons of safety.
3. Brainstorm ideas for the building project such as a house, a school, a hospital, or a store. Write the suggestions on the easel and discuss the options together. When ready, decide on one idea for the class architectural project.
4. Pretend to be the architect and tell the children that they are now the owners. Explain that they are going to meet with you and decide what their building needs. Ask questions such as: *What materials do you want your building to be made out of? What shape do you want it to be? How many rooms do you need? How many doors and windows? How many bathrooms? Do you need any closets? How big should they be?* Write the children's questions and responses on the easel so that they can refer to them when working on their class blueprints.
5. Conclude by talking about next week's lessons. The children will be making a floor plan and building their special class architectural project.

Materials
- Paper
- Easel
- Tape
- Marker

Preparation
- Tape a length of paper to the easel. Prepare another piece of paper to attach when ready.

Teaching Tip
- Be sure to ask specific questions that are related to the needs of your building. For example, if the children are going to make a school, ask if there should be a sink in every classroom. If they are building a hospital, ask how big the doorways should be so that the beds can fit through.

©McGraw-Hill Children's Publishing

WPH39005 *Preschool Projects: Construction*

Class Elevation Plan

Activity 16

Objective
The children will learn that they need a well-thought out plan before they start to build.

Directions
1. Gather the children together and talk about last week's activities. Explain that today they are going to make an elevation plan for their special building project.
2. Review the list of questions and responses with the children. Ask if they can think of anything else to include in their building. Add these items to the list.
3. Tape a piece of paper onto the easel next to the list of responses and explain that the children are now going to help draw their class elevation plan (exterior view). Review the sample elevation plans as necessary.
4. Taking turns, the children can help draw the front of the building including the roof, windows, and doors. They can use lead pencils, rulers, and T-squares to make straight lines and angles, refering to the list as needed.
5. Encourage the children to take turns adding color to their elevation plan. They may include trees, grass, vehicles, and multicultural people on the plan.
6. After the elevation blueprint is finished, review the plan together. Display or roll it up to store in a safe place for future activities.
7. Conclude by talking about tomorrow's lesson. The children will be drawing a floor plan for their special class building project.

Materials
- List of questions and responses (See page 26.)
- Paper
- Easel
- Elevation plans (samples)
- Pencils (lead and colored)
- Rulers
- T-squares

Preparation
- Tape the list of questions and responses to the easel.
- Prepare a length of paper to attach to the easel when needed.

Teaching Tips
- Sample blueprints may be homemade or donated by architects, builders, or parents.
- Some children may like to draw side and rear views of the elevation plan to share with their classmates.

Activity 17

Class Floor Plan

Objective
The children will learn that they need a well-thought out plan before they start to build.

Directions
1. Gather the children together and talk about yesterday's activity. Explain that today they are going to make a floor plan for their special building project.
2. Review the list of questions and responses and the class elevation plan with the children. Ask if they can think of anything else to include in their building. Add these items to the list.
3. Tape another piece of paper onto the easel and explain that the children are now going to help draw their class floor plan. Review the sample blueprints as necessary.
4. Draw the main outside shape of the building (probably a square, rectangle, or circle) with black marker. Use the ruler and T-square to make horizontal and vertical lines.
5. Invite the children to take turns drawing items inside the main shape to represent doors, windows, sinks, cabinets, or other details. They can use lead pencils, rulers, and T-squares to make straight lines and angles refering to the list as needed.
6. As the children draw each detail on the floor plan, label the part correctly (e.g., door, window). Then number each wall on the floor plan (from outside to inside) in preparation for the class model.
7. After the blueprint is finished, review the plan together. Display or roll it up to store in a safe place for future activities.
8. Conclude by talking about tomorrow's lesson. The children will be building a class model of their special architectural project.

Materials
- List of questions and responses (See page 26.)
- Class elevation plan (See page 27.)
- Easel
- Tape
- Paper
- Floor plans (sample)
- Markers (black)
- Pencils
- Rulers
- T-squares

Preparation
- Tape the list of questions and responses and class elevation plan to the easel.
- Prepare a length of paper to attach to the easel when needed.

Teaching Tip
- Sample blueprints may be homemade or donated by architects, builders, or parents.

Class Model

Activity 18

Objective
The children will make a model of their building so they have a clear idea of what their final project will look like.

Directions
1. Gather the children together and talk about yesterday's activity. Explain that today they are going to use their floor plan to make a small model of their building.
2. Invite the children to stand around a big table making sure they can all see the floor plan and construction materials. Have the children take turns naming or telling what they can see on this blueprint.
3. Give each child a piece of precut cardboard or balsa wood. Help the children look for and identify the numbers on their pieces.
4. Ask the two children with numbers 1 and 2 on their "walls" to come forward. They can match their numbers with the ones on the floor-plan base and glue their walls together.
5. Working in numerical order, the other children can take their turn gluing their pieces to the model (from outside to inside).
6. After assembling all the walls, add dollhouse furniture and trees to complete the scenario. Place the model in a learning center for the children to gently examine, explore, and discuss.
7. Conclude by talking about tomorrow's lesson. The children will be pretending to prepare the site for construction.

Materials
- Class floor plan (See page 28.)
- Balsa wood or cardboard
- Craft knife (adult use only)
- Pencils or markers
- White glue
- Glue brushes
- Dollhouse furniture and trees

Preparation
- Make a base for the model by copying the floor plan onto balsa wood or cardboard.
- Cut pieces of balsa wood or cardboard into correct sizes for the model walls. Number the pieces in order, according to the floor plan. (See page 28.)
- On a large table, place the pieces of balsa wood or cardboard, the base, glue, and brushes for the children.

Teaching Tip
- Invite the children to compare and contrast the class model with their floor and elevation blueprints.

WPH39005 *Preschool Projects: Construction*

©McGraw-Hill Children's Publishing

Activity 19

Getting the Site Ready

Objective
The children will learn that a site needs to be prepared for construction often by big machinery.

Directions
1. Gather the children together and talk about yesterday's activity. Explain that today they will be pretending to get the site ready for construction.
2. Share the following information with the children.
   - Since structures need to be built on solid, level ground, contractors must prepare the site for construction.
   - The leveling, digging, and moving of the site is often done by big machinery. Bulldozers move away the large rocks; loaders push dirt around to level the surface; backhoes dig big holes into which the foundations can be poured; cement mixers mix and pour cement for foundations.
3. Read *Tonka: Big Book of Trucks* to the children and focus on common construction vehicles that are often used to get a site ready (bulldozers, loaders, backhoes, cement mixers). Have volunteers move the toy construction vehicles about to show how they can prepare the sandbox or tray of sand for building.
4. Ask the children to pretend that they are driving big machines such as bulldozers, loaders, backhoes, and cement mixers to get the classroom "site" ready for construction.
5. When the children have finished, they can step back from their site to see how well their crew worked together. They can discuss the success of their efforts and take a rest.
6. Conclude by talking about tomorrow's lesson. The children will be building the final class architectural project.

Materials
- Toy construction vehicles (e.g., bulldozers, loaders, backhoes, cement mixers)
- Sandbox or tray of sand
- *Tonka: Big Book of Trucks* by Patricia Relf

Preparation
- Place the toy construction vehicles in the sandbox or tray of sand.
- Clear an area of the classroom or perform the activity outside.

Teaching Tips
- The children can also pretend to be the actual machines.
- Provide real or make-believe hard hats, fluorescent orange vests, and boots (from parents or tradespeople) for your construction workers to wear.
- Encourage the children to use toy machinery to prepare building sites in the learning centers. (See pages 6–8.)
- Arrange for the children to visit a construction site to view big machinery at work. (See page 9.)

Final Project

Activity 20

Objective
The children will experience being part of a "real" building project.

Directions
1. Gather the children together and talk about this month's activities. Review the class blueprints and model, then explain that today they will be working on their final project.
2. Share the following information with the children.
   - Different crews of tradespeople work together or in turn to pour concrete foundations and put up roof and wall frames. They finish the exterior walls and roof, install doors and windows, and run electrical, plumbing, gas, and other lines. They also complete and paint the interior walls and ceilings and install cabinets, tubs, flooring, lighting, appliances, and other fixtures.
3. Working together, help the children put cardboard boxes in place to form the frame, then tie or tape them together. Add more boxes and pieces of cardboard for walls. Cut holes for windows and doors as necessary. Be sure the children check the blueprints!
4. After the frame and walls have been built, the children can work on the inside. They can use simple tools and pretend to be plumbers, electricians, and carpenters. (E.g., plumbers may use wrenches to install pretend washbasins, electricians can put up plastic wires, carpenters can put up pieces of wood for shelves). The children can also decorate the building with paint, markers, and fabric.
5. Celebrate the completion of this project with a ribbon-cutting ceremony! The children can take turns sitting, standing, and playing inside their building.
6. Discuss the finished project together and invite parents and other classes to come and view the children's architectural endeavors.

Materials
- Class blueprints (See pages 27 and 28.)
- Easel
- Tape
- Class model (See page 29.)
- Building materials (e.g., large pieces of cardboard, cardboard boxes, bed sheets, balsa wood, fabric, plastic wire and piping, string, ribbon, paint, markers, glue)
- Tools (e.g., hammers, nails, wrenches)

Preparation
- Tape the class blueprints to the easel and display the model.
- Clear an area of the classroom and place building materials and tools nearby.

Teaching Tips
- Remind the children that the building process takes a lot longer in real life.

WPH39005 *Preschool Projects: Construction* 31 ©McGraw-Hill Children's Publishing

Bibliography

Barton, Byron. *Building a House* (Mulberry/Greenwillow, 1981).

Devlin, Harry. *What Kind of a House Is That?* (Parents' Magazine Press, 1969).

Dorros, Arthur. *This Is My House* (Scholastic, 1992).

Florian, Douglas. *A Carpenter* (Greenwillow, 1991).

Gaughenbaugh, Michael and Herbert Camburn. *Old House, New House: A Child's Exploration of American Architectural Styles* (The Preservation Press, 1993).

Gibbons, Gail. *Up Goes the Skyscraper!* (Four Winds/Macmillan, 1986).

Glenn, Patricia Brown. *Under Every Roof: A Kids' Style and Field Guide to the Architecture of American Houses* (The Preservation Press, 1993).

Goldreich, Gloria and Esther. *What Can She Be? An Architect* (Lothrop, Lee and Shepard, 1974).

Isaacson, Philip M. *Round Buildings, Square Buildings, and Buildings That Wiggle Like a Fish* (Alfred A. Knopf, 1988).

Jessop, Joanne. *The X-Ray Picture Book of Big Buildings of the Modern World* (Franklin Watts, 1994).

Knapp, Brian. *The World Around Us: Homes of the World and the Way People Live* (Grolier Educational Corporation, 1994).

Relf, Patricia. *Tonka: Big Book of Trucks* (Scholastic, 1996).

Seltzer, Isadore. *The House I Live in: At Home in America* (Macmillan, 1992).

Stevenson, Neil. *Annotated Guides: Architecture* (Dorling Kindersley, 1997).

Thorne-Thomsen, Kathleen. *Frank Lloyd Wright for Kids* (Chicago Review Press, 1994).

Wilkinson, Philip. *Amazing Buildings* (Dorling Kindersley, 1993).

Williams, Gene B. *Be Your Own Architect* (Tab Books/McGraw-Hill, 1990).

Video and Internet Resources

Bigham, Vicki Smith and George. *The Prentice Hall Directory of Online Education Resources* (Prentice Hall, 1998).

Polly, Jean Armour. *The Internet Kids and Family Yellow Pages* (Osborne/McGraw Hill, 2000).